FOCUS ON ELEMENTARY

CHEMISTRY

Rebecca W. Keller, PhD

REAL SCIENCE 4 Kids

Cover design: David Keller
Opening page illustrations: David Keller
Text illustrations: Janet Moneymaker, Rebecca W. Keller, PhD

Focus On Elementary Chemistry Student Textbook (softcover)
ISBN 978-1-936114-56-6

Published by Gravitas Publications, Inc.
http://www.gravitaspublications.com

GRAVITAS
PUBLICATIONS

Chapter 1   Atoms

1.1 Atoms

Have you ever wondered if the Moon is really made of green cheese?

Have you ever thought the clouds might be made of cotton candy?

Have you ever
wanted to
know
what
makes
carrots
orange...

or peas
green?

Have you ever wondered why
brussels sprouts couldn't taste
more like sweet cherries, or
asparagus taste more like candy
canes?

Everything around us has a
different shape or flavor or color,
because everything around us
was designed with different atoms
put together in different ways.

Atoms are very small things we can't see with only our eyes.

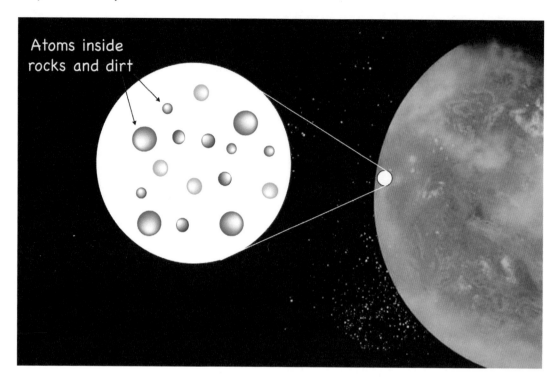

The Moon is not really made of green cheese. It is made of atoms that are found in rocks and dirt.

Clouds are not made of cotton candy, but of atoms found in air and water.

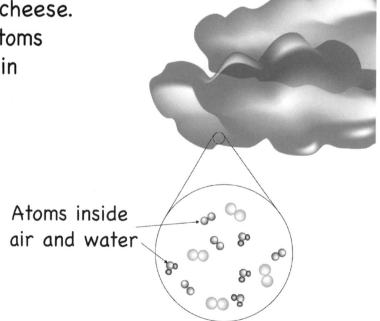

Carrots are orange because their atoms are arranged in a way that makes them orange. Peas are green because their atoms are arranged in a way that makes them green. Brussels sprouts and asparagus don't taste sweet like cherries or candy canes because the atoms inside brussels sprouts and asparagus are not arranged in a way that makes them sweet.

1.2 Different Atoms

There are over 100 different atoms.
Carbon, oxygen, and nitrogen are the names of a few different atoms.

Atoms are very small; they are so small that you can't see them with only your eyes. Even though we can't see atoms we can draw them as little balls.

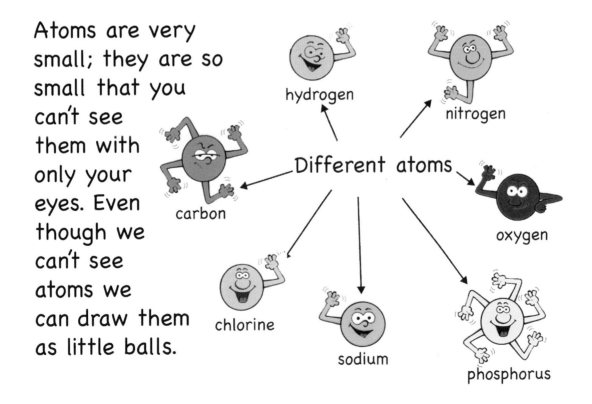

1.3 Atoms Stick Together

Atoms can be by themselves, or they can hook to other atoms to make molecules. We will learn more about molecules in Chapter 2.

Atoms can stick together in many different ways. The different ways that atoms stick to other atoms make things different from each other.

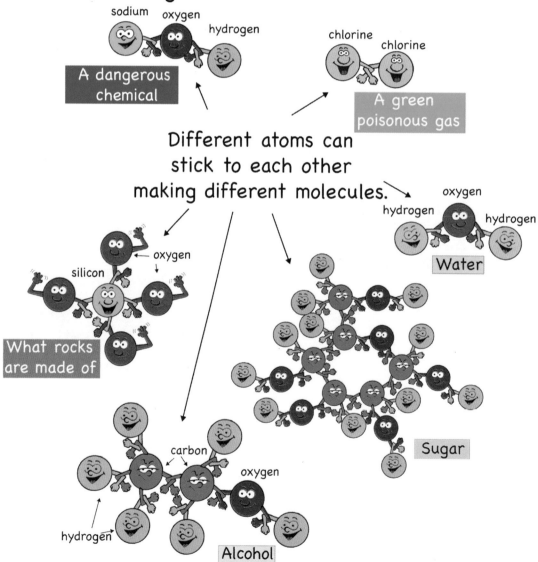

sodium   oxygen
hydrogen
A dangerous chemical

chlorine
chlorine
A green poisonous gas

Different atoms can stick to each other making different molecules.

oxygen
hydrogen
hydrogen
Water

silicon   oxygen
What rocks are made of

carbon
oxygen
hydrogen
Alcohol

Sugar

1.4 Making Observations

How do we know everything is made of atoms if we can't see them with our eyes?

We know about atoms because of the way things look and the way things behave. To learn about atoms, scientists make very careful observations about the world around them.

But scientists aren't the only ones who make observations. Everyone does! Making observations just means looking at things and wondering about them. *You* make observations all the time.

For example, when you see an ant crawling on the ground, you are making an observation. You might ask yourself, "What color is the ant?"

"How many legs does it have?"

"Does it crawl in a straight line, or does it wander?" All of these questions, and many others, can be answered by making careful observations.

Observations are a very important part of science, because it takes careful observations to discover new things. You might think you know what something looks like, but when you observe it carefully, you might find something new!

1.5 Summary

○ Everything is made of atoms.

○ Atoms are very small things we can't see using only our eyes.

○ Atoms can stick to other atoms to make things taste different, feel different, look different, or smell different.

○ Making careful observations helps scientists and *you* make new discoveries about the world.

Chapter 2 Molecules

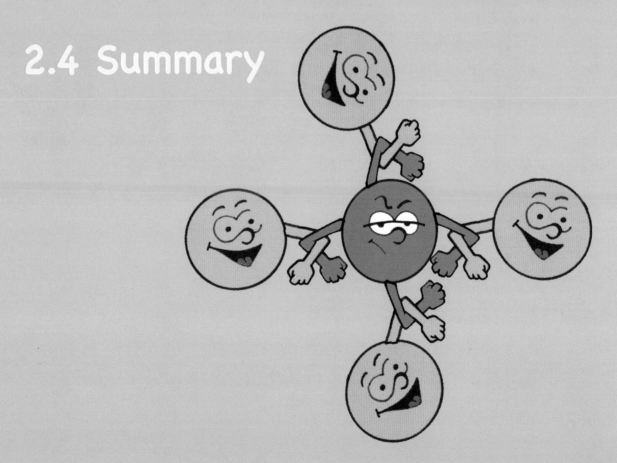

2.1 Atoms Are Building Blocks

In the last chapter we saw that everything is made of atoms. The Moon is made of atoms. The clouds are made of atoms. Carrots, peas, brussels sprouts, and asparagus are all made of atoms.

But how do atoms make so many different things?

If we think of atoms as building blocks, we can understand how atoms make so many different things.

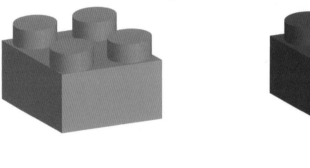

2.2 Atoms Form Molecules

Just as plastic building blocks are designed to hook to each other to make toy buildings, toy cars, or toy boats, atoms are designed to hook to each other to make molecules.

Sometimes only a couple of atoms hook together to make a molecule.

For example, table salt is made of just two atoms: sodium and chlorine.

Water is made of three atoms—two hydrogen atoms and one oxygen atom.

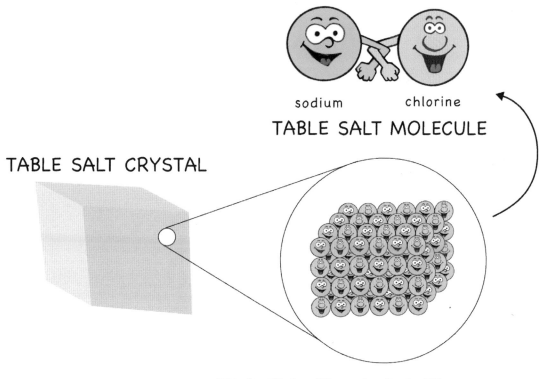

sodium   chlorine

TABLE SALT MOLECULE

TABLE SALT CRYSTAL

TABLE SALT

Other molecules are made of more atoms.

Table sugar is made of 41 atoms.

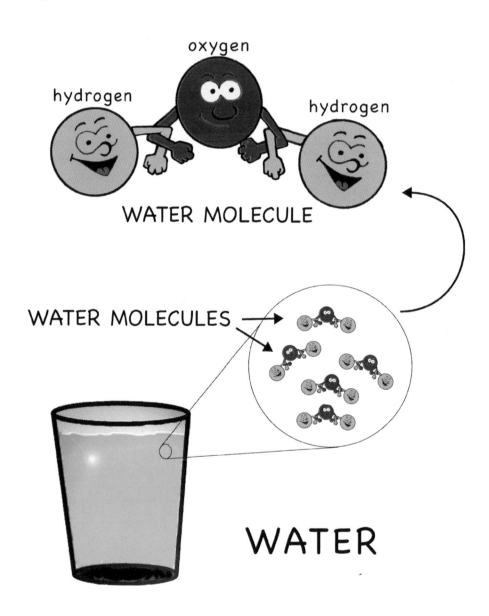

WATER MOLECULE

WATER MOLECULES

WATER

Vegetable oil is made of 170 atoms or more.

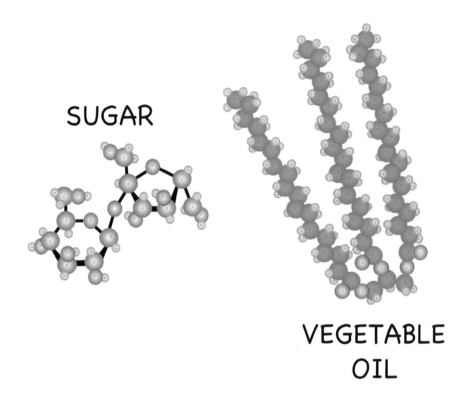

SUGAR

VEGETABLE
OIL

Some of the molecules in your body are made of thousands and thousands of atoms.

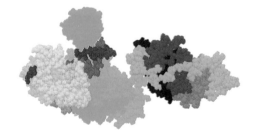

A molecule in your body
with thousands
of atoms.

2.3 Atoms Follow Rules

Atoms hook to other atoms in special ways. They follow rules!

For example, hydrogen atoms can only hook to one other atom. Hydrogen cannot hook to two atoms or three atoms.

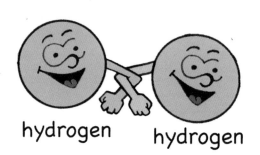

hydrogen   hydrogen

Oxygen cannot hook to more than two atoms.

oxygen

nitrogen

Nitrogen can hook to one, two, or three atoms. It cannot hook to more than three atoms.

Carbon can hook to one, two, three, or four atoms. Carbon cannot hook to more than four atoms.

Many different shapes and sizes of molecules can be made with atoms. However, atoms always obey rules to make molecules. These rules mean that table salt will be table salt, and sugar will be sugar!

2.4 Summary

- Atoms hook to other atoms to make molecules.

- Molecules can have lots of different shapes and sizes.

- Atoms have to obey rules to make molecules.

Chapter 3 Molecules Meet

3.1 When Molecules Meet

In the last chapter we saw that atoms hook together to make molecules. We also found that atoms must obey rules. Each atom hooks to other atoms in its own way. But what happens when one molecule meets another molecule? What do they do? Do they change or do they stay the same?

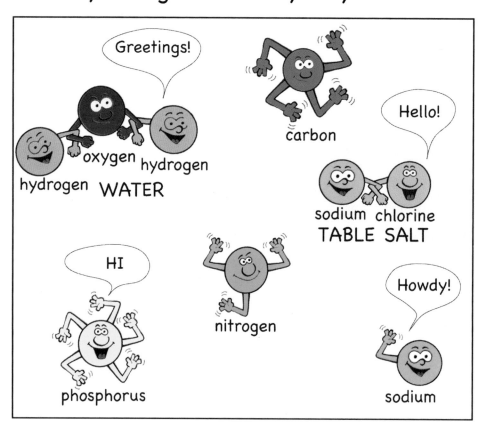

Sometimes when one molecule meets another molecule, they react. This means that there is a change in the way the atoms of the molecules are hooked together.

3.2 Molecules Switch Atoms

Sometimes molecules react by switching atoms [exchanging partners]. In the example below, two molecules meet and trade atoms. As a result, two new molecules are made.

1 Two molecules meet.

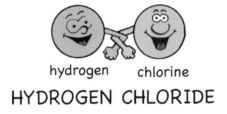

hydrogen   chlorine

HYDROGEN CHLORIDE

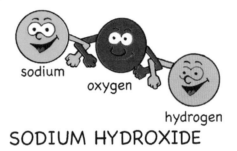

sodium

oxygen

hydrogen

SODIUM HYDROXIDE

2 The hydrogen atom and the sodium atom switch places.

3 Two new molecules are made.

sodium   chlorine

SODIUM CHLORIDE
(table salt)

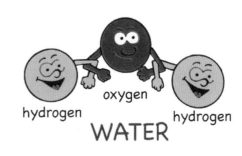

hydrogen

oxygen

hydrogen

WATER

3.3 Molecules Join Together

Sometimes when molecules meet, they join together. In this example, two chlorine atoms in a chlorine gas molecule meet two sodium atoms. The chlorine atoms and the sodium atoms combine to make table salt!

❶ Two chlorine atoms [in a chlorine gas molecule] meeting two sodium atoms.

CHLORINE GAS MOLECULE   SODIUM ATOMS

❷ The chlorine atoms join the sodium atoms to make two sodium chloride molecules [table salt].

TABLE SALT

3.4 Molecules Break Apart

Sometimes molecules might simply break apart to form new molecules. In the next example, two

water molecules break apart and then join together to make oxygen gas and hydrogen gas.

1 Two water molecules.

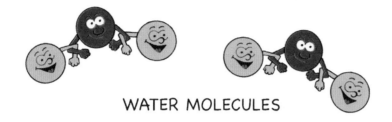

WATER MOLECULES

2 The water molecules break apart.

3 New molecules are made.

HYDROGEN GAS   OXYGEN GAS

All of these examples show how molecules and atoms react with one another. It is important to realize that in every reaction atoms are neither created nor destroyed. Atoms can rearrange themselves and change places to make new molecules, but atoms never simply appear or disappear.

3.5 Reactions Are Everywhere

When atoms switch places, join together, or leave, a chemical reaction has occurred.

There are lots of chemical reactions. They go on all the time and all around us. For example, the gasoline inside a car reacts with oxygen to provide energy for the car to move.

Reactions occur when you bake bread or cook an egg.

Even the red rust you find on your metal shovel if you leave it in the rain is caused by a chemical reaction.

Reactions occur inside your body. When you eat a piece of cheese or drink a glass of milk, reactions occur inside your mouth. These reactions help break down the food molecules into smaller pieces. Inside your stomach there are strong molecules that break your food down still further. Even when you breathe, reactions inside your lungs help oxygen get into your

blood so it can be carried through your body. Reactions are everywhere.

3.6 Reactions Follow Rules

Reactions also have to follow rules. Not every molecule will react with every other molecule or atom. Some molecules won't react at all. For example, the noble gases, such as neon, helium, and argon usually don't react with any other molecules.

Some molecules react with lots of other molecules. Water will react with many things. Water will even start a fire in some reactions!

3.7 We Can See Reactions

Often we can observe something happening if a reaction is occurring. Sometimes we can see bubbles. Sometimes we might see little particles form that look like sand. Sometimes the glass might change temperature in our hands if a reaction is happening. There could also be fire, an explosion, or a color change.

All of these observations tell us that a reaction may be happening.

Bubbles

Fire

Particles
(like sand)

3.8 Summary

○ When atoms get rearranged, a chemical reaction has occurred.

○ Atoms can switch places, join together, or separate from each other.

○ In a chemical reaction, atoms rearrange but are never created or destroyed.

○ Reactions occur everywhere.

○ We may be seeing a reaction take place when there are bubbles, color changes, or temperature changes.

Chapter 4 Acids and Bases

4.1 Special Molecules

In the last chapter we saw that when atoms or molecules meet, a reaction sometimes happens. We saw how atoms can trade places, join together, or separate from each other during a reaction. In this chapter we will look at some special kinds of molecules called acids and bases. Acids always react with bases, and bases always react with acids.

4.2 Acids and Bases Are Different

Have you ever noticed that when you bite a lemon it tastes sour and makes your cheeks pucker?

Have you ever tasted mineral water or baking soda water? They are not sour like a lemon. They are bitter or salty.

Have you ever noticed that soap is very slippery in your fingers, but lemon juice and vinegar are not?

The molecules inside a lemon are different from the molecules inside baking soda water or mineral water. Lemons have molecules in them called acids. It is the acid in lemons that gives them their sour taste.

Baking soda water and soap have molecules in them that are called bases. Bases often make things feel slippery or taste bitter.

4.3 H and OH Groups

We have seen how acids and bases are *different* from each other in many ways. This is because a base is a different kind of molecule than an acid. Acids and bases are different because they have different atom groups.

oxygen

hydrogen

OH group

hydrogen

A base has an OH group [say "O" "H" group]. An OH group is just an oxygen atom and a hydrogen atom together.

H group

Most common acids have an H group [say "H" group]. An H group is just a hydrogen atom.

We can see in the next picture that sodium hydroxide (a base) has an OH group and hydrogen chloride (an acid) has an H group.

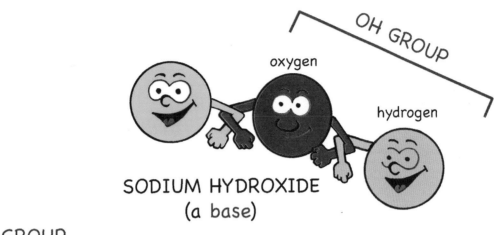

SODIUM HYDROXIDE
(a base)

HYDROGEN CHLORIDE
(an acid)

4.4 Both Are Important

Both acids and bases are very important. They are needed in lots of very useful reactions. You have a strong acid inside your stomach to break down your food. Without the acid in our stomachs, we could not digest our food.

Acids are also found in batteries, lemons, oranges, grapes, and even soda pop.

These things have acids in them.

Bases are found in lots of cleaners like window cleaner, bathroom cleaner, and soap. They are also found in some foods like bananas or dates. Bases are even used to make your stomach feel better! We'll see why in the next chapter.

These things
have bases
in them.

4.5 Summary

○ Acids taste sour.

○ Bases taste bitter and are slippery.

○ Acids have an H group and bases have an OH group.

○ Acids and bases are found everywhere — in batteries, in your stomach, in household cleaners, and even in bananas or lemons!

Chapter 5 Acids and Bases React

5.1 When Acids and Bases Meet

In the last chapter we learned about two different kinds of molecules: acids and bases. We saw that acids and bases are found in lots of different things. Acids are in batteries, lemons, and even soda pop. Bases are in soap, window cleaner, and even bananas!

What happens when acids and bases meet? As we learned in Chapter 3, when molecules meet sometimes they react. Do acids and bases react when they meet?

1. An acid and a base meet.

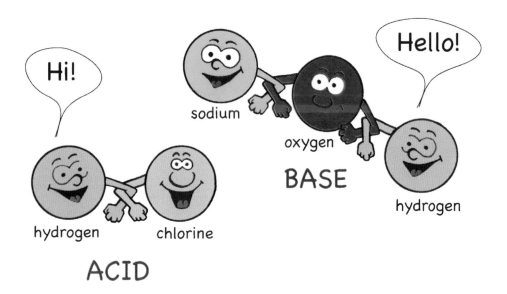

5.2 Acid-Base Reactions

In fact they do! Acids and bases make a special kind of reaction called an acid-base reaction. Like other reactions we saw in Chapter 3, the atoms in the acid exchange with the atoms in the base when they meet.

After they meet, some atoms leave their molecules.

2. The atoms leave their molecules.

Next, the atoms that left go to the other molecules and "make new friends."

3. The atoms make new friends.

Now, two new molecules have been made. The new molecules for this reaction are water and table salt (sodium chloride).

4. Two new molecules are made.

The acids and bases are no longer acids and bases. When they react, they become other kinds of molecules, such as salt and water.

5.3 Important Acid-Base Reactions

Acid-base reactions are very important. For example, your stomach has acid in it. This acid is necessary for digesting your food. Sometimes

there is too much acid. When this happens, your stomach hurts. The medicine your mom or dad may give you is a base. It reacts with the acid in your stomach, turning it into a salt and water. That makes your stomach stop hurting.

5.4 Observing Acid-Base Reactions

Sometimes we can *see* an acid-base reaction. Some acids and bases give off heat or explode when they react. Other times we cannot tell when an acid-base reaction happens. When we can't see an acid-base reaction, we can put something into the reaction that shows us that the reaction is taking place. This "something" is called an indicator because it indicates, or tells us, something is happening.

Indicators

We use indicators all the time. Stop lights indicate when we can go or when we should stop. When we turn on the oven, an indicator tells us when it is hot enough. A thermometer is an

indicator. It can tell when your body has a fever. Indicators are used all the time. They are also used in chemistry.

An acid-base indicator tells us whether we have an acid or a base. There are different kinds of acid-base indicators. A simple acid-base indicator is red cabbage juice! Red cabbage juice turns pink with acids and green with bases.

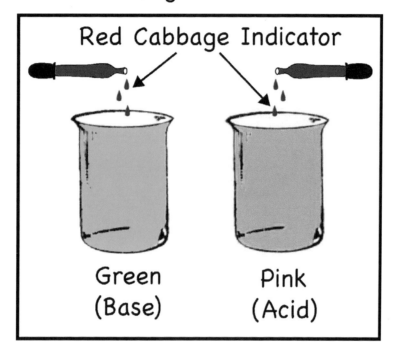

5.5 Summary

○ Acids and bases react with each other in acid-base reactions.

○ Many acid-base reactions make salts and water.

○ Acid-base indicators can tell us when we have an acid or when we have a base.

Chapter 6 Mixtures

6.1 Mixing

Have you ever put water and sand together in a pail? What did you get? A mud pie maybe!

Have you ever made a real pie, like lemon pie? If you have, you

probably added eggs and flour, some table salt and oil, and maybe some water. What happened when you added all these things together? You probably mixed them with a spoon or a mixer.

In either case, what you ended up with is a mixture.

A mixture of sand and water or a mixture of eggs, oil, lemon, and water—both mud pies and lemon pies are mixtures.

6.2 Mixtures

You can make a mixture of blocks and rocks. You can make a mixture of rocks and sand. You can make a mixture of sugar and cinnamon and put it on your toast! All of these are called mixtures because all of these are made of more than one thing *mixed* together.

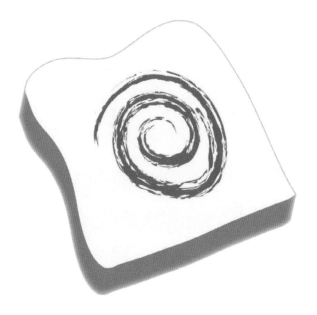

6.3 Some Mixtures Dissolve

Have you ever wondered why table salt disappears in water, but sand does not? Have you ever noticed that sugar disappears in water but not in oil or butter? When table salt or sugar disappear in water, we say they dissolve.

Some things will dissolve in water and some things will not dissolve. What makes some things dissolve and other things not dissolve?

Table salt crystals dissolve (break apart) in water.

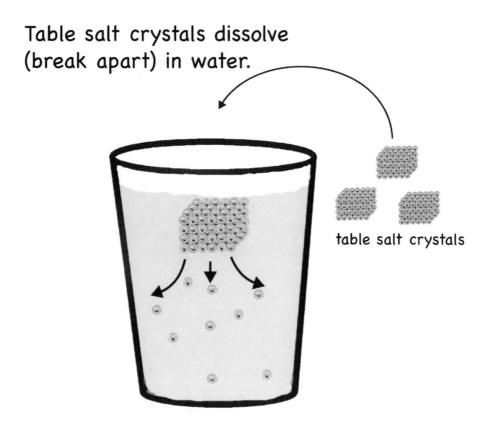

table salt crystals

6.4 Dissolving

As with everything else, it's the molecules in table salt and sugar that determine whether or not they will dissolve.

Molecules have to follow rules for dissolving or not dissolving, just like they have to follow rules for reacting or not reacting.

The main "rule" for dissolving is:

Like dissolves like.

This means, for example, that molecules that are "like" water *will* dissolve in water and molecules that are "not like" water *will not* dissolve in water.

This doesn't mean that the molecules have to be identical or *exactly* alike, they just need to have a few things in common.

For example, what makes some molecules "like" water? We saw in Chapter 4 that acid molecules have an H group and bases have an OH group. If we look carefully at water, we see that it has BOTH an OH group and an H group! This is one of the things that makes water very special.

H group

OH group

oxygen

hydrogen   hydrogen

WATER

It is the OH group that makes molecules dissolve in water. Bases that have OH groups are "like" water and will dissolve in water. Other molecules, like alcohol, which is not a base but still has an OH group, will also dissolve in water.

Sugar is "like" water because sugar also has OH groups. Can you count how many OH groups sugar has?

1. Alcohol, sugar, and sodium hydroxide are "like" water — they have OH groups.

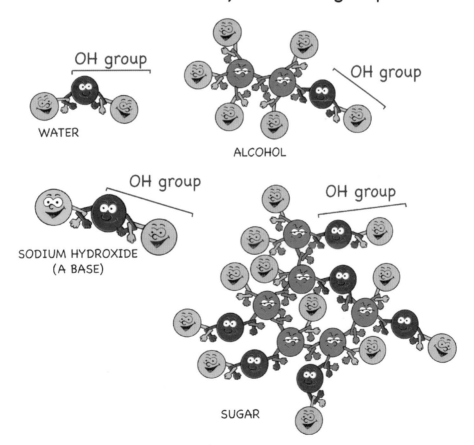

It's not just OH groups that make some things dissolve in water. For example, salt doesn't have OH groups like sugar, alcohol, and bases do, but salt dissolves in water. Salt dissolves in water because the water molecules break the salt molecules into pieces that mix with water.

2. Salt will dissolve in water.

First the salt breaks apart...

...and then the salt atoms mix with the water molecules.

Oil, grease, and butter are not like water, so none of these will dissolve in water. Look carefully at the drawing that illustrates the type of molecule found in oil, grease, and butter. Can you tell why it is not like water?

3. Oil, grease, and butter are not like water.

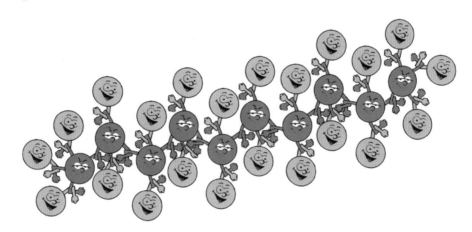

TYPE OF MOLECULE FOUND IN OIL, GREASE, AND BUTTER

6.5 Soap

Soap makes things like butter and grease "dissolve" in water. Soap can do this because the molecules that make up soap are a little like water and a little like oil.

In a mixture of oil, soap, and water, the oily part of soap will dissolve in the oil and the watery part of soap will dissolve in the water.

4. Soap has an "oil-like" part and a "water-like" part.

oil-like part of soap

water-like part of soap

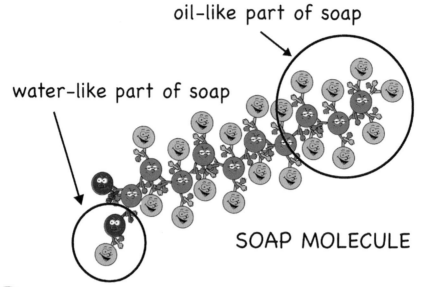

SOAP MOLECULE

5. The oil-like part of soap dissolves in the oil, and the water-like part dissolves in the water.

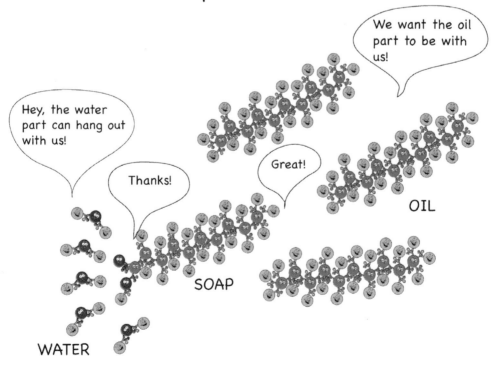

WATER, SOAP, AND OIL MIXTURE

Because the oil dissolves in the oily part of soap, and the watery part of soap dissolves in the water, a small droplet of oil and soap forms. In this way, the oil is "trapped" by the soap and water inside this little droplet.

6. Droplet of oil molecules and soap surrounded by water molecules.

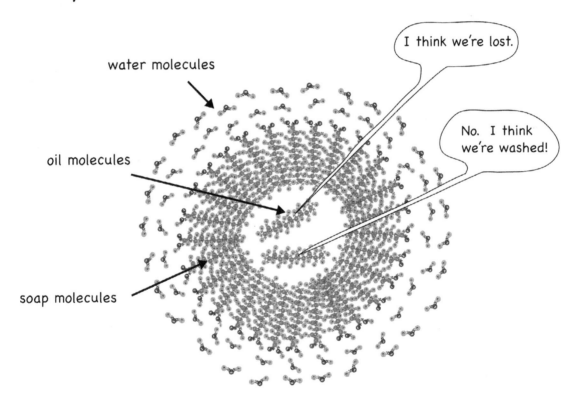

This droplet can then be washed away by the water. This is how soap washes the grease off your hands!

6.6 Summary

○ Mud pies and lemon pies are mixtures.

○ A mixture is anything that has more than two types of items in it.

○ Some mixtures dissolve. Others do not.

○ Dissolving depends on the kind of molecules in the mixture. Molecules that are "like" each other dissolve. Molecules that are "not like" each other will not dissolve.

○ Soap is like both water and oil. This means that soap can make oil "dissolve" in water.

Chapter 7 Un-mixing

7.1 Un-mixing

In the last chapter we learned about mixtures, but how do we get things that are mixed to "un-mix?" Can we get the water and sand to "un-mix" from a mud pie? Can we get the eggs, sugar, water, and lemons to "un-mix" from a lemon pie?

Try to think of ways to "un-mix" a mud pie. What if you let the mud pie bake in the sun? What happens to the water? What happens to the sand?

7.2 Evaporation

You may know that the water "disappears" from the mud pie, and the sand stays behind. We say that the water has evaporated. Evaporation is one way to "un-mix," or separate, mixtures that have water in them.

What happens if we leave the lemon pie to bake in the sun? Will the lemon pie "un-mix?" The water will evaporate, but what happens to  the eggs, sugar, and lemons? They do not evaporate. In fact, they stay behind and we have a not-so-tasty lemon mess!

7.3 Sorting By Hand

Sometimes we can "un-mix" things and sometimes we cannot. The mud pie can be "un-mixed," by the sun, but the lemon pie cannot.

Large things are usually easy to "un-mix." Even though, when your mom tells you to clean your room, a large pile of toys may seem impossible to "un-mix"—with some work, it can be done.

All of the toys are easy to pick up because they are large. They can be picked up with your hands and put into separate bins or boxes.

7.4   Using Tools

What about a pile of sand? The sand cannot be easily picked up because each piece of sand is very small. It would take hours to pick up all of the sand by hand!

Fortunately a tool can be used. Can you think of a tool for picking up sand and "un-mixing" it from your carpet?

That's right—a vacuum cleaner! A vacuum cleaner can be used as a tool for "un-mixing."

In fact, tools are used all the time to "un-mix" things that are hard to "un-mix" with your hands. For example, sieves or colanders are used to separate hot spaghetti or hot potatoes from boiling water.

7.5 Using "Tricks"

There are other tools and other ways to "un-mix" mixtures of small things. What about molecules that you can't even see? Are there ways to separate molecules?

There are! In fact, scientists use a trick called chromatography to separate molecules. Using chromatography, you can un-mix many different kinds of molecules.

One type of chromatography is called paper chromatography. With paper chromatography, a piece of paper is used to separate small things like molecules. You can use paper chromatography to separate the small molecules that are in ink or dye, or even the molecules found in a colored flower!

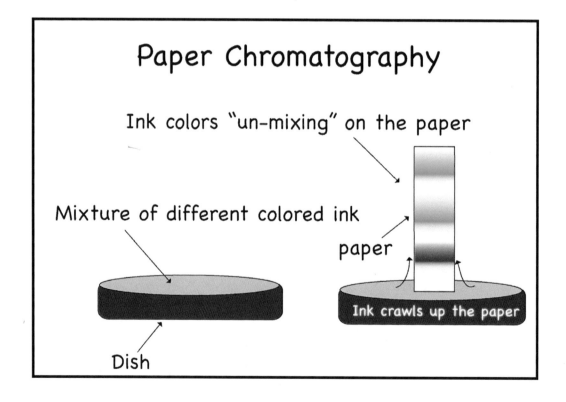

Paper Chromatography

Ink colors "un-mixing" on the paper

Mixture of different colored ink

paper

Ink crawls up the paper

Dish

7.6 Summary

○ The sand and water in a mud pie can be "un-mixed" by evaporating away the water.

○ Some mixtures, such as lemon pies, cannot be easily "un-mixed."

○ Mixtures of large things are easier to "un-mix" than mixtures of smaller things.

○ Tools, like vacuum cleaners or sieves, can be used to "un-mix" some mixtures.

○ A trick called chromatography can be used to separate molecules.

Chapter 8 Food and Taste

8.1 Tasty Molecules

Now you know why vinegar and lemons taste sour—they're acids! And why mineral water and soda water taste bitter—they're bases! Have you ever wondered why salt tastes salty or sugar tastes sweet?

We have learned that everything around us is made of atoms. The food we eat is made of atoms, but not all of the food we eat tastes the same. Why? Different molecules in different foods make foods taste different.

1. Acid molecules found in soda pop...

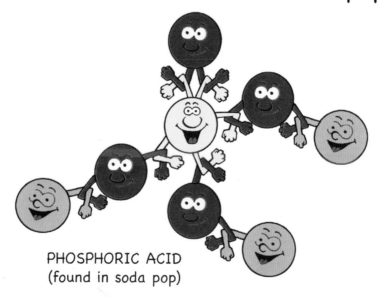

PHOSPHORIC ACID
(found in soda pop)

...make soda pop taste sour*.

(*soda pop also has lots and lots of sugar in it, so it also tastes very sweet.)

2. Acid molecules in vinegar...

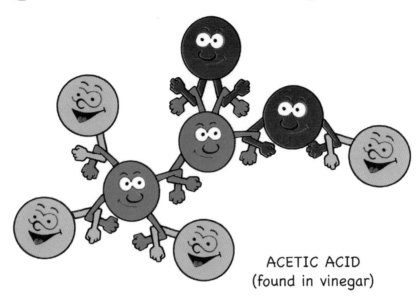

ACETIC ACID
(found in vinegar)

... make vinegar taste sour.

We have already seen that sour foods often have an acid in them. Lemons, vinegar, and grapefruit have acid in them. When you eat foods with acid in them, your tongue tells your brain "sour."

Foods that have salt in them taste salty. Salt molecules look very different from acid molecules. Remember that table salt has a sodium atom and a chlorine atom hooked together. When you eat foods with salt in them, your tongue tells your brain "salty."

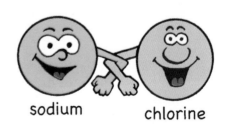

sodium   chlorine

SODIUM CHLORIDE
(table salt)

8.2 The Amazing Tongue

Your tongue is designed to tell your brain what kind of molecules are in your food. It can sense acids, bases, salt, sugar, and many other molecules.

For example, foods that taste sweet have sugar in them. A sugar molecule looks different from a salt molecule or an acid molecule. It is larger, has more atoms in it, and some of the atoms are hooked together in a ring. When you eat a piece of candy, your tongue tells your brain "sweet."

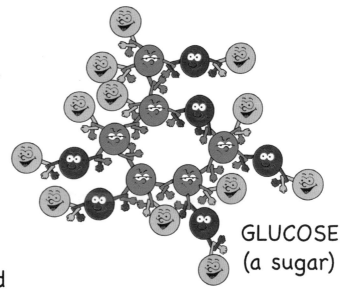

GLUCOSE
(a sugar)

Your tongue can tell the difference between a salt molecule, an acid molecule, and a sugar molecule.

Your tongue is a remarkable indicator.

Remember from Chapter 5 that an indicator senses and tells us something. There are lots of man-made indicators, like thermometers and stop lights, but there are none that are as intricately designed as your own tongue!

8.3 Large Tasty Molecules

Not all molecules are tasted by your tongue. For example, a raw potato doesn't exactly taste sweet, salty, bitter, or sour, but something in between. In fact, a potato is made mostly of sugar molecules, but your tongue can't tell. Your tongue can't taste the sugar molecules in a raw potato because they are hooked together in long chains. A long chain of sugar molecules is called a carbohydrate. Cooking breaks the long chains of molecules into loose, single sugar molecules, and your tongue can then taste them.

one glucose molecule

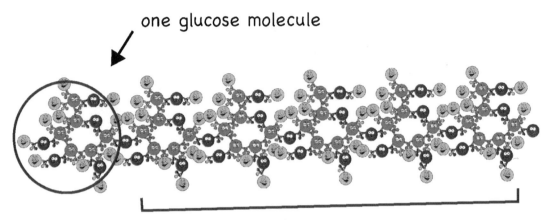

chain of glucose molecules

CARBOHYDRATE

Many different foods have carbohydrates. Bread, pasta, potatoes, and many fruits have carbohydrates in them. Carbohydrates are important molecules for your body. Because they are made of sugar molecules, they provide the energy your body needs to ride a bike or climb a tree!

8.4 Summary

- Foods taste different because foods are made of different molecules.

- Your tongue is an amazing indicator that can tell the difference between salts, acids, bases, and sugars.

- Carbohydrates are long chains of sugar molecules.

- Your tongue can detect sugar molecules in foods that contain carbohydrates if the carbohydrates are broken apart.

Chapter 9 Molecular Chains

9.1 Chains of Molecules

In the last chapter we saw how sugar molecules hook together to form a long chain. We found out that long chains of sugar molecules are called carbohydrates.

There are other kinds of long chains not made of sugar molecules. Long chains can be made out of many different kinds of molecules. In general, long chains of molecules are called polymers.

Polymer

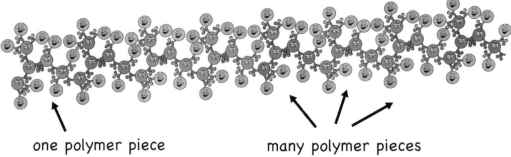

one polymer piece   many polymer pieces

Polymers are everywhere! Almost anywhere you look, you can find polymers. Your clothes, your toys, your food, and your hair are all made of polymers!

9.2 Different Polymers

Plastics are polymers. Your toy car and parts of your dad's car are made of polymers.

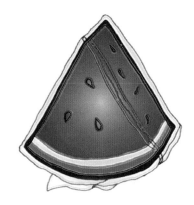

The plastic wrap you put over your food is made of polymers.

The plastic cup you drink from and the plastic pen you write with are made of polymers.

Rubber is also a polymer. The rubber hose outside in the garden, a rubber ball, and the rubber boots you wear on rainy days are all made of polymers.

Styrofoam is a polymer. The packing peanuts that come with your new chemistry kit are made of polymers. The styrofoam cup that holds your dad's coffee or your lemonade is made of polymers.

Your clothing is also made of polymers. Cotton fibers, nylon fibers, polyester, and wool are all polymers. Polymers are everywhere!

9.3 Polymers Can Change

How can polymers make so many different things if they are all just long molecules?

Because different polymers have different properties (like being sticky or stiff), many different things can be made with polymers.

Sometimes long chains of polymers slide up and down next to each other. Polymers like this, such as glue and natural rubber, can be sticky.

Sometimes long chains of polymers are hooked together and can't slide up and down next to each other. Polymers like this can be hard or stiff and not sticky.

It is possible to change the properties of polymers by using chemicals or heat. For example, egg whites are made of polymers. When you cook an egg, you change the properties of the polymers inside the egg. The egg whites change from a clear, sticky liquid to a firm, white solid when you cook them. The polymers inside the egg whites have changed their properties because of heat!

9.4 Summary

○ Polymers are long chains of smaller molecules all hooked together.

○ Polymers are everywhere! Plastics, rubber, styrofoam, and clothing are made of polymers.

○ Different polymers give objects different properties. Some polymers make things soft, and some polymers make things hard or stiff.

○ You can change the properties of a polymer by use of chemical reactions or heat.

Chapter 10 Molecules in Your Body

10.1 Special Polymers

In the last chapter we looked at long chains of molecules called polymers. We learned that plastics, rubber, clothing, and food are all made of polymers.

Did you know that many of the molecules inside your body are also polymers? There are many different polymers inside your body. We will learn about two of them.

One very special kind of polymer is called a protein.

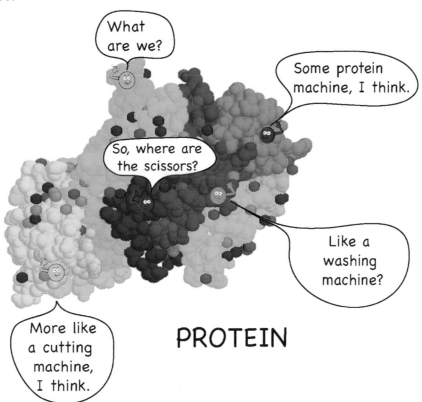

PROTEIN

A protein is a long chain of small molecules hooked together and often folded up into a special shape. It is the special shape of this folded chain of molecules that helps proteins do amazing things.

10.2 Proteins—Tiny Machines

Proteins are tiny machines inside your body that perform incredible tasks. In fact, proteins do almost all of the work inside your body.

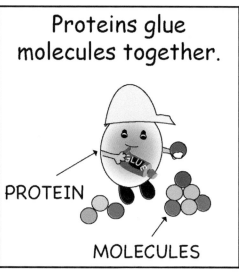
Proteins glue molecules together.

PROTEIN

MOLECULES

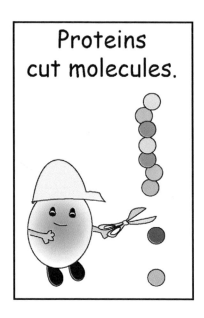
Proteins cut molecules.

Some proteins glue other molecules together.

Some proteins cut other molecules.

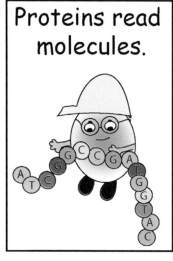
Proteins read molecules.

Some proteins copy other molecules and some proteins "read" other molecules.

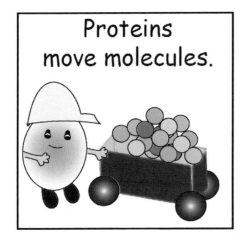

Proteins
move molecules.

And some proteins move other proteins or molecules from place to place.

Proteins do an amazing number of different jobs inside your body.

10.3 DNA — A Blueprint

One of the molecules that proteins read, cut, paste, and carry is called DNA. DNA is also a polymer.

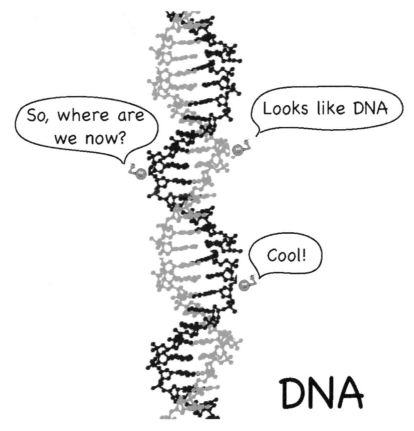

DNA

DNA is a very special molecule. It is not just any ordinary polymer. DNA is special because DNA carries your genetic code. A genetic code is like a set of instructions. Your genetic code determines if you will have brown hair or blonde hair. Your genetic code tells whether you will have green eyes or blue eyes, or whether you will have light skin or dark skin. The genetic code carried by your DNA is essentially the blueprint for your body.

Everybody has a different and unique genetic code, or blueprint. You get your blueprint from your parents, and they got their blueprint from their parents. Your parent's parents got their blueprint from their parents, and so on. Your blueprint tells what you will look like or how tall you may grow, but your blueprint doesn't tell everything about you.

Where you live, what you eat, what you do, and even what you think makes you unique and not like any other person who ever lived or who ever will live! Even identical twins who have identical DNA are different from each other. You are more than just your DNA. You are uniquely designed in every way.

10.4 Summary

○ There are polymers in our bodies.

○ Some polymers are called proteins. Proteins are tiny machines that glue, cut, copy, and carry molecules in your body.

○ Some polymers are called DNA. DNA carries the genetic code.

○ Your body is an amazing design of large and small molecules, polymers, and genetic information. You are uniquely designed.

Made in the USA
Columbia, SC
20 June 2018